Choice Theory:

A Simple Introduction

Also by K.H. Erickson

Simple Introductions

Choice Theory
Financial Economics
Game Theory
Game Theory for Business
Investment Appraisal
Microeconomics

Choice Theory:

A Simple Introduction

K.H. Erickson

© 2013 K.H. Erickson

All rights reserved.

No part of this publication may be reproduced, stored in or introduced into a retrieval system, or transmitted in any form or by any means, including electronic, mechanical, photocopying, recording or otherwise, without the prior permission of the author.

Contents

1 Introduction   7
2 Expected Value Theory   10
2.1 Expected Value   10
2.2 Two Envelopes Problem   16
2.3 St. Petersburg Paradox   19
3 Expected Utility Theory   21
3.1 Expected Utility   21
3.2 Ellsberg Paradox   32
3.3 Allais Paradox   35
3.4 Preference Reversal Phenomenon   38
4 Prospect Theory   40
4.1 Original Prospect Theory   40
4.2 Cumulative Prospect Theory   48
4.3 Third Generation Prospect Theory   52
5 Risky Games   58
5.1 Certain and Uncertain Games   58
5.2 Minimax Strategies   63
6 Auction Theory   70
6.1 Types of Auction   70
6.2 Private Value Auctions and Revenue Equivalence   73
6.3 Common Value Auctions and the Winner's Curse   77

7 Voting 81
7.1 Voter Preferences and the Condorcet Winner 81
7.2 Condorcet Voting Cycles 91
7.3 Playing the System 95
Bibliography 98

1 Introduction

Choice Theory examines the processes people engage in during decision-making, and specifically what they do when faced with conditions of risk or uncertainty. In conditions of certain outcomes the choice is easy and they'll always select the highest payoff. But when faced with a range of possible outcomes a number of factors come into play, including the possible odds and value of each outcome, and this book examines how individuals try to navigate this uncertainty in search of better results.

The subject of choice theory is divided into two parts; choice under risk, and choice under uncertainty. Choice under uncertainty looks into situations with unknown probabilities of events occurring, and may represent a world where an individual has limited power among a larger group. Choice under risk assumes only unknown outcomes due to a range of possible results, and outcome probabilities are fully known and can be acted upon as desired by an individual.

This book focuses first on choice under risk, and looks at the factors influencing human decision-making when probabilities are fixed but outcomes are varied. Common sense may suggest that individuals would go for the outcome they expect will offer the highest value, and

Expected Value Theory is explained in depth with examples used to calculate the value of an uncertain outcome. Well known criticisms of the model are then examined, with the Two Envelopes Problem and St. Petersburg Paradox challenging the idea that choice is based on expected value.

Focus then turns to Expected Utility Theory and the idea that an individual's own risk attitude can affect their valuation of outcomes, as risk aversion and risk seeking behaviour are explained, along with the steps used to find an individual's utility function. But there is evidence of violations of the model's core assumptions and three examples are put forward here, with the Ellsberg Paradox, Allais Paradox, and Preference Reversal Phenomenon suggesting that individual choice under uncertainty is not as predictable as may be expected.

An alternative model is put forward with Prospect Theory, which looks at gains and losses differently and centres on the idea of a reference point from which prospects are evaluated. The prospect model has evolved over time, and the three different versions are all presented with graphical representations of their key features, and some conclusions drawn as to their use in predicting human behaviour when facing risky choices.

The second part of the book turns to choice under conditions of uncertainty, with three separate areas addressed where the probability of outcomes is not necessarily known, as other people enter the equation and individual choice is affected by group choice. Risk attitudes are applied to the competitive game theory model of human interaction, as people turn to minimax or maximin strategies to manage the uncertainty brought by co-dependent interactions with others, and secure a better payoff.

Auction Theory examines the situation where a choice must be made to outmanoeuvre competing bidders without paying over the odds. The revenue equivalence theorem shows how a bidder can ensure the auction type doesn't affect his winning bid, and a strategy to avoid the Winner's Curse of overpaying for an object is also discussed.

Voting strategies in a world where an individual is outnumbered is the focus in the final part of the book. The idea of a Condorcet Winner that can't lose the vote against any alternative is examined, along with voting cycles where voters don't have a clear collective preference. Small changes to the choice of alternatives put before voters can make all of the difference to the result, and a situation where the government manipulates the outcome is put forward, along with the reaction of the voters who may choose to play the system in response.

2 Expected Value Theory

2.1 Expected Value

Imagine a generous friend coming to you with an appealing proposition. He's willing to give you free money, and all you have to do is choose which of the two amounts on offer you'd prefer; £10 or £20. This decision is easy and everyone would prefer the £20. Those who aren't sure can replace the pounds sterling with a choice between $10 and $20, or €10 and €20 etc. Whatever your currency the choice is between two amounts, where one is twice the size of the other. That is choice under conditions of certainty, where the outcome is fixed and known and an individual simply has to pick the option with the highest return.

There is no need to look into theory when faced with certainty, as the process is predictable and each selection leads to a guaranteed result. But operating under conditions of risk is a more common situation, where there are a range of possible outcomes for every choice. Those faced with this scenario want to know what to expect, and whether the outcome is likely to be positive or negative.

Consider the same wealthy acquaintance as before returning with a different offer: he will toss a coin to decide whether you get £10 or £20, or alternatively he'll give you £12 right now. This choice requires some thought, and if you call the coin toss correctly you'll get £20 and be glad to have turned down the £12 up front, but if you're wrong then it's only £10, filling you with regret over missing out on the £12 offer and losing two pounds. With a fair coin the odds of heads or tails is equal, and the amount won is purely down to random chance.

A model based on expected value can help an individual make a well-informed decision when faced with multiple outcomes. Expected Value (EV) Theory calculates the level of wealth linked to each outcome and weighs the probability of each result against its value. This makes the two options comparable and facilitates a smart choice. The probability of each outcome is simply the percentage chance of it occurring, which can be written as either a fraction or decimal, and the value of each outcome is the money prize on offer. EV theory argues that people will select the choice they believe offers the maximum return, as they always seek to maximize their wealth level.

The following diagram gives a visual representation of expected value theory, and an increase in the probability of gaining the prize or in the size of the outcome will increase the expected value, while a decrease in either factor has the opposite effect on expected value.

Expected value

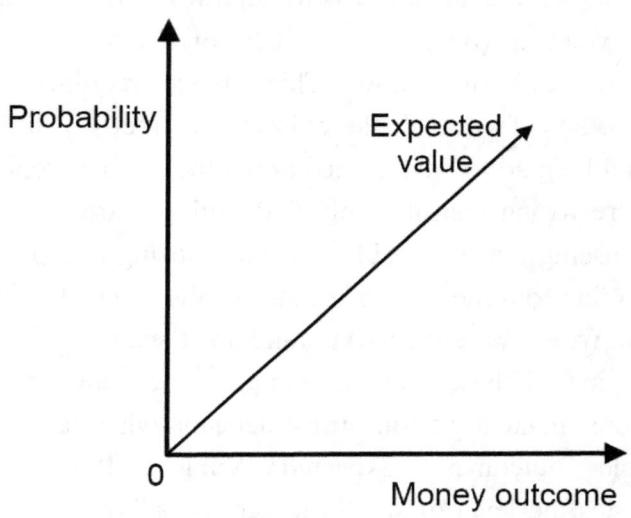

The generous friend has already given the money payoffs from his offer: £10, £20 or £12. If we take the offer of £12 up front, not concerned that a higher payoff was possible and wanting to avoid depending on random chance, then the expected value would be 100%*£12:

$$EV = 1*12 = \underline{£12}$$

Those willing to place their future wealth in the hands of a coin toss would face 50/50 or 1/2 odds of calling heads and tails correctly, assuming a fair coin, and that would give 50% odds of both £10 and £20:

$$EV = (1/2*10) + (1/2*20) = 5 + 10 = \underline{£15}$$

With this information a well-informed choice can be made, and the logical answer is to go with the coin toss. It offers a higher expected value of £15, compared to being given £12 with no questions asked.

On the face of it expected value theory is appealing and it feels intuitively correct. People use EV theory every day people to weight up the pros and cons of an action, and decide whether they're happy with the range of possible outcomes before proceeding. Anyone who buys a product they intend to sell for a profit has used EV theory, and they assign a certain probability of it making a good price and a certain probability to it losing them money, finding the overall expected value acceptable to proceed with the purchase. Meanwhile those who look at the product and then back away calculated different probabilities, and believed that the chance of it losing them money was higher and the odds of it being worthwhile were lower, making it not worth their time and effort.

Two traders can examine stock to sell wholesale and come up with two different valuations. Trader A sees the stock for sale at £4 a unit, and believes there's a 25% or one in four chance that each unit of the stock won't sell, and it'll be left as useless to him with a return of £0. But he's confident that there's a 75% or three in four chance that he can sell each unit at a price of £6 to his customers.

Unfortunately there isn't the option to buy a smaller amount of the stock for a greater likelihood of selling it all, and the wholesale supplier will only offer the price of £4 a unit for the set large order. Any smaller an order and the price will jump up to £6 a unit which clearly wouldn't offer a profit.

For it to be worth his while to buy the product, trader A must believe he will get more for each unit in expected value than what they would cost him:

$$EV = (1/4*0) + (3/4*6) = 0 + 4.5 = \underline{£4.50}$$

The opportunity to buy wholesale stock is one worth taking for trader A. They will only cost him £4 a unit, but according to his calculations can be sold at an average return of £4.50 each. That's a £0.50 profit per unit.

The second buyer interested in wholesale stock is trader B, and he also finds the stock for sale at £4 a unit if a large order is made. He believes there's a 20% or a one in five risk that each unit won't be sold, and will be left as useless stock with a return of £0 to him, and an 80% or odds of four in five that he can get a price of £4.80 per unit. For it to be worthwhile buying the product trader B must receive more in EV than his unit cost:

$$EV = (1/5*0) + (4/5*4.8) = 0 + 3.84 = \underline{£3.84}$$

It isn't worth trader B buying the wholesale stock, even at the deal price of £4 per unit. He predicts that overall he'd only be able to gain an average unit return of £3.84, and that means he's losing £0.16 for every unit bought. That's clearly something he'd want to avoid.

As the example shows, EV theory allows for different valuations of outcomes. It shows a process that businesses and individuals follow to make a living, and allows for different calculations to be made with the same basic data. People use expected value every day and place their futures in its hands.

But not everyone would take part in a gamble and risk of ending up worse than they started, even if there was a huge prize on offer at fair odds. Anyone who had very little and needed every note they had couldn't accept the risk. A hungry man who needed it for an immediate meal, or a traveller whose car had broken down and wanted the money to pay for a bus home, would be unlikely to risk ending up with nothing for a 50/50 shot at a big cash payout. This is one of the possible flaws of expected value theory, and there are others that tear the theory apart completely.

2.2 Two Envelopes Problem

The two envelopes problem looks at a theoretical situation where a test participant is stuck in a never-ending cycle, as they move back and forward in an attempt to chase greater returns. Two identical envelopes each contain a sum of money, with one holding twice as much as the other. An individual can pick one envelope and keep the money inside, but before they open their selection they're given the chance to switch envelopes. The test begins and a person picks up an envelope and is offered the chance to swap it. They know that the other envelope that they don't have either contains double (if they have the less valuable envelope), or half (if they have the more valuable envelope) of what they hold in their hand.

If the envelope being held were to contain £10, for example, then the other one either has £5 or £20 inside. If it's the former they shouldn't switch, but if it's the latter then they should. No-one can know for sure, but with only two envelopes there's a 50/50 shot at each, and an participant would need to decide whether they should hold on to the (assumed) £10, or take the 50/50 risk for a chance of either £5 or £20. This is where expected value (EV) theory would be used. For the envelope held with the (assumed) £10:

$$EV = 1*10 = \underline{£10}$$

And for the other envelope there's a 50/50 chance of each outcome:

$$EV = (1/2*5) + (1/2*20) = 2.5 + 10 = \underline{£12.50}$$

With the EV of the other envelope coming in at £12.50, more than the £10 assumed to be in the current envelope held, it makes sense to switch. But once the switch is made the individual will wonder whether they made the right choice. Still assuming that the EV of the envelope now in hand is £12.50, they know that the other envelope must have an EV of either double this (if the more valuable of the two envelopes) or half of it (if the less valuable envelope). That means the other envelope's EV is now either £6.25 or £25, with a 50/50 fair chance of each:

$$EV = (1/2*6.25) + (1/2*25) = 3.125 + 12.50 = \underline{£15.63}$$

The envelope that the individual originally picked but switched away from has an EV of £15.63, which is greater than the EV of £12.50 for the current envelope. If the test participant selecting between the two envelopes was making decisions based on the maximum expected value, then he would switch away from an envelope and then back again, and this cycle would repeat forever. No matter how many times the envelopes were switched the EV of the other envelope would always be higher than the current one. The possibility that it contains double the money sees it match the value of the current envelope, while still getting half in a worse case scenario gives it an overall higher value. Yet clearly no real person in this situation would switch the envelopes forever, which contradicts EV theory and its belief that people seek to maximize expected value.

2.3 St. Petersburg Paradox

The St. Petersburg paradox looks at a theoretical lottery of infinite value, according to the calculations made by expected value theory, but that no-one would pay a large amount of money to play. In the 18th century David Bernoulli thought up the lottery where a fair coin is tossed at each stage of the game, with the pot to win starting at one pound sterling and doubling each and every time a head is thrown. As soon as a tail appears the game ends and the player wins whatever is in the pot, but until that point the game continues and the pot rises with each head. If a tail were thrown on the first toss of the fair coin the player would win £1, not until the second gives £2, only on the third £4, the fourth £8 and so on.

The expected value of the game is decided only by the probability and outcome linked to a tail appearing on the coin, and when a head appears the game simply continues and there is no payoff at that point. If a fair coin is used where there's equal chance of a head or tail, then the expected value of the lottery is:

$$(1/2*1) + (1/4*2) + (1/8*4) + (1/16*8) + (1/32*16) +\ldots$$
$$= 0.5 + 0.5 + 0.5 + 0.5 + 0.5 + 0.5 +\ldots$$
$$= \infty$$

In the first round there's a one in two that a tail will be thrown to end the game, and by the second round a one in four overall chance, as the total odds of seeing a single tail halve each turn. In the first stage the payoff for throwing a tail is 1 and each further stage sees the payoff double. With the probability of the single tail result required to get the value of the pot halving each turn, and the prize for it doubling, these two effects cancel each other out in terms of expected value. Every single round past round one is essentially just a replication of that stage, and its payoff of 0.5. No matter how unlikely it may be for a tail to not appear until the one thousandth toss for example, that thousandth round still offers a payoff of 0.5 in the lottery.

With the lottery set to continue indefinitely until a tail is finally thrown, the expected value of the game is an indefinite summation of 0.5 payoffs, and it's worth an infinite amount of money. Yet no-one on earth would be willing to trade their life savings for a chance to play this game, even though expected value theory would suggest that is a good deal. People would be willing to pay a handful of pounds sterling at most, and this challenges the usefulness of expected value as a model to predict choice under risk. The predictions the model makes simply don't match the reality of human behaviour, and it has limited use as a tool. A new model is required that doesn't show what people should do but what they actually do in reality.

3 Expected Utility Theory

3.1 Expected Utility

The problem with expected value theory is that it only calculates actual wealth values, based on the monetary outcome on offer and the odds of receiving it, and assumes that people act on this factor and this alone.

Imagine it's lunchtime at an office workplace and a worker is considering what to get. A catering service is provided by the office but each item comes at a price, and the man's primary concern is to get something that will fill him up to avoid him being distracted from work. Suddenly a trolley is pushed by and the man sees that there are three different options today: a meat sandwich, noodles, and a salad. The woman pushing the trolley expects the office worker to go for the noodle box, as they've been popular and it's the lowest cost choice. But the man instead pays over the odds for the meat sandwich, as he's confident that the lean meat will keep his hunger at bay better than anything else. Expected value had no impact on his decision here as he actually went for the lowest expected value item, in terms of using up his own money, but he followed his expected utility to make his choice.

Even though individuals may claim that their only motivation is money, in reality they only want the money for what it would be spent on. This means they will not only evaluate a lottery based on the prize on offer, but also on how much they personally value it. It's not the level of wealth that matters to people but the utility of wealth and the satisfaction derived from a certain level of it, and this involves Expected Utility (EU) Theory. A graph can be plotted to show an individual's utility of wealth or EU function, where wealth levels are plotted against an individual's utility of wealth and the relationship gives their expected utility.

To find an individual's EU function they can be asked to give monetary valuations for different outcomes. Theoretical questions may give results that wouldn't hold up in real life situations, and researchers often prefer to run experiments where participants are presented with a series of outcomes or gambles:

How much would they pay for a 50/50 fair coin toss to win either £10 or £0; £20 or £0; £30 or £0; £40 or £0; and £50 or £0?

If they owned the gamble already and were selling the opportunity to others, what is the lowest price they'd demand for the 50/50 fair coin toss gamble to win £10 or £0; £20 or £0 gamble, etc.?

With the results from these questions aggregated over a number of individual research participants, a graph could be plotted with a point for the price people were willing to pay (their utility of wealth) against the actual value of the gamble (wealth or expected value). The actual or expected value (EV) of the gambles here is easy to calculate:

EV= (1/2*10) + (1/2*0) = 5 + 0 = £5;
EV= (1/2*20) + (1/2*0) = 10 + 0 = £10;
EV= (1/2*30) + (1/2*0) = 15 + 0 = £15;
EV= (1/2*40) + (1/2*0) = 20 + 0 = £20;
EV= (1/2*50) + (1/2*0) = 25 + 0 = £25.

This can then be linked with the people's valuations.

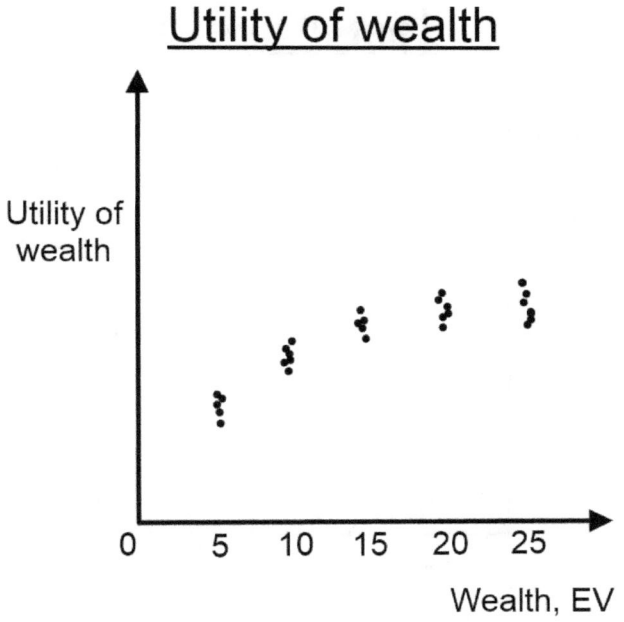

The five collections of dots in the diagram represent various utility values that individual participants may give to the five different wealth or EV outcomes. For example, one person may have been willing to pay £5 for the 50/50 gamble to win either £10 or £0 (£5 expected value), while others may have only valued the gamble at £4.50 at most, or £4.20, etc. For the 50/50 gamble to win either £50 or £0 (£25 expected value), all participants may be willing to pay far less as the graph above suggests, and one person may have paid £16 for it, another would offer £15 at most, while one person may demand £12.50 to sell it, etc. With the data points above a best fit or trend line can be plotted.

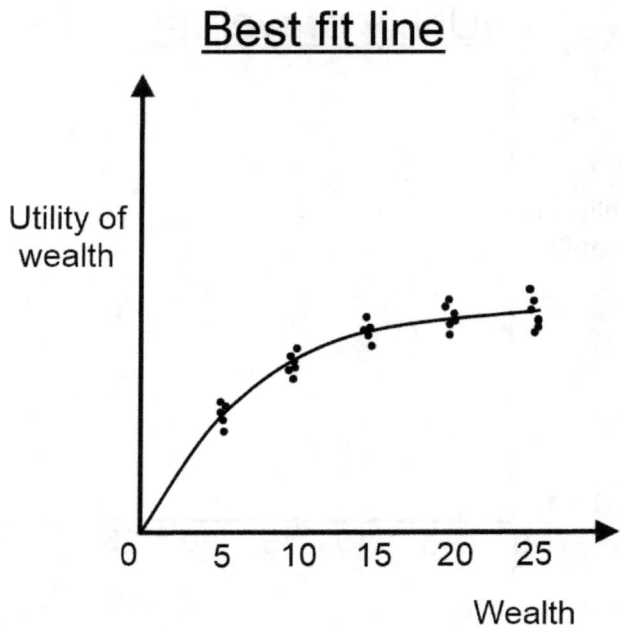

With the best fit line plotted the individual points can be removed, and the line that remains is the expected utility of wealth or expected utility (EU) function. This offers a guide as to how changes in wealth can affect people's utility levels.

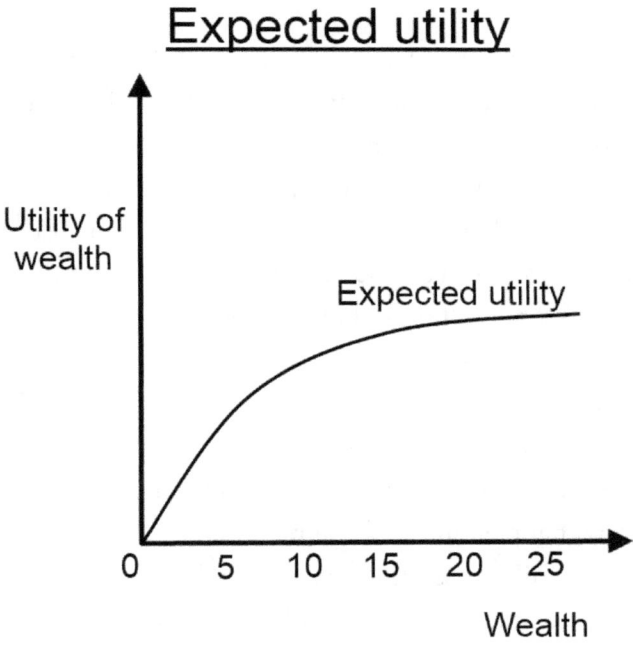

The EU curve has now been extended beyond a wealth or expected value level of £25, and based on the curvature at earlier points the relationship between wealth and utility can be extrapolated at higher wealth levels. This graph is purely theoretical and various shapes of curve may be found in reality, but once the function curve is found an equation could be calculated to find the exact relationship between utility (U) and wealth (W). The previous curve may exhibit a relationship akin to:

$$U = \sqrt{(W)} - 5$$

Irrespective of what the exact relationship is, the information is of vital importance to policy makers throughout society. Employers want to know the relationship between employee salary and happy productive workers for a cost effective but efficient workforce, while governments wonder how much they can levy in essential taxes before facing a revolt from disgruntled citizens. Looking beyond wealth in purely monetary terms and instead seeing it in terms of any desirable good, advertisers are also desperate for the information, to market their products more efficiently based on what customers value most.

In the following diagram the individual's expected utility is 'risk neutral' as shown by the straight line, where utility of wealth has a fixed relationship with wealth. In this situation expected value theory as explained earlier would be correct, and a risk neutral individual doesn't change his attitude with higher levels of wealth but values the outcome purely based on the facts of the numbers on offer. However, risk neutrality is rare among people, and other types of risk attitude are more common.

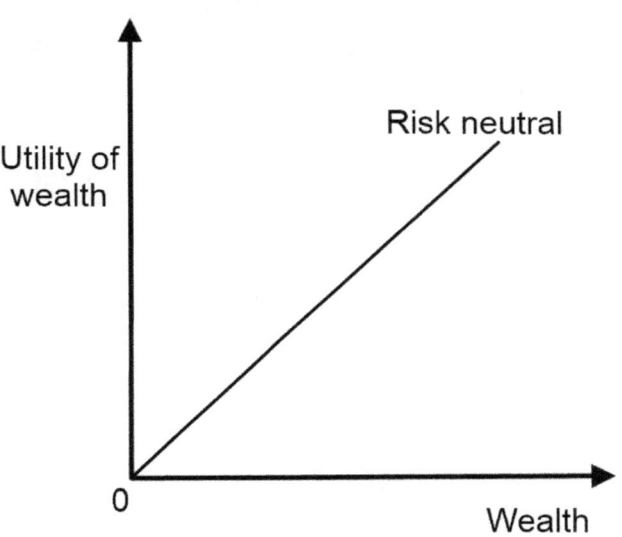

An individual with this EU function could be someone whose wealth is all put toward discretionary spending, and the utility of money is simply what it can buy.

Example EU functions for an individual with risk neutral preferences, where W is wealth or expected value:

$$EU = 0.5W$$
$$EU = W$$
$$EU = 3W$$

The curved EU function that follows shows a different attitude to risk, and demonstrates 'risk aversion' is a person. After a minimum threshold of wealth has been met, where the curve first appears in sight, the individual gains a disproportionately low amount of utility as wealth rises. This is visible with the steep curve that becomes flatter as wealth grows further.

A risk averse utility function

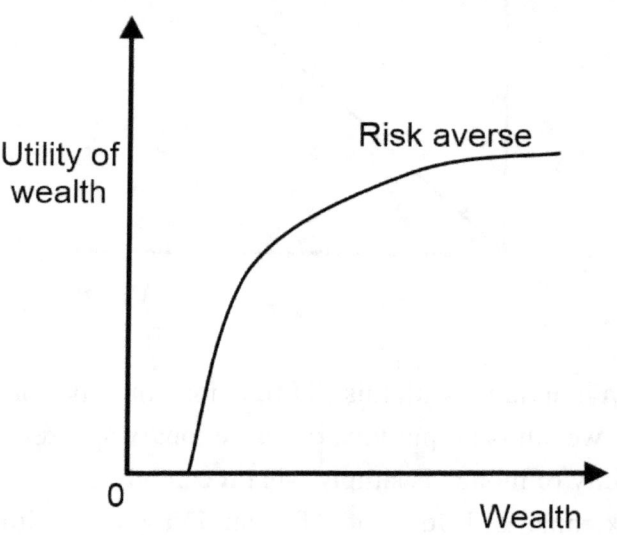

This means that above a minimum threshold the individual will have more to lose if their wealth is taken away, and they will be wary of taking risks, although this trait will decline as they become wealthier. This person may be someone whose money is put toward achieving a basic standard of living. Wealth may offer no utility until a certain amount is attained to allow a subsistence level of income, but after this threshold the wealth secures a certain highly-satisfying lifestyle which the individual doesn't want to give up. And as wealth grows further it only adds increasingly unnecessary luxury goods, and there is no great desire to risk lower levels of wealth to gain the chance of higher wealth. A risk averse person will suffer more from a loss than they benefit from an equivalent size gain.

Example EU functions for an individual with risk averse preferences:

$$EU = \log (W)$$
$$EU = \sqrt{(W)} - 1$$

The third type of utility function is 'risk seeking' or risk loving where wealth adds little to utility until high levels are reached, and the expected utility curve stays almost horizontal until it steepens with higher wealth.

A risk seeking person won't be affected by low level wealth changes, and may be happy to spend small amounts of money without feeling that anything is lost. But with high wealth levels there is a large jump in utility to suggest that such a person seeks expensive indicators of high status, and will risk low levels of wealth to try and get it. A risk loving person benefits more from a gain than they suffer from an equivalent size loss.

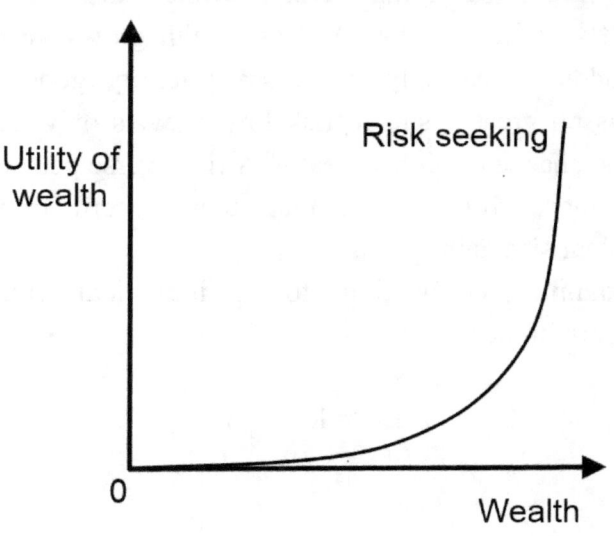

Example EU functions for an individual with risk seeking or risk loving preferences:

$$EU = W^2$$
$$EU = 2W^2$$
$$EU = W^3$$

Expected utility theory implies that individuals are rational and seek to maximize their expected utility, while the model's ability to represent various attitudes to wealth levels and risk appears to resolve the shortcomings of expected value theory. But EU theory depends on certain conditions being met, called the von Neumann-Morgenstern axioms. Individuals must have a complete, consistent, continuous, independent and equivalent set of preferences, with a preference for higher probabilities of outcomes over lower ones. This means that individuals must have an EU function that looks vaguely like one or a combination of the three images just shown. There should be a clear set of visible preferences and there no gaps in the EU function, where their utility level is unknown. Nor should it curve back on itself, which would mean two levels of utility for the same level of wealth, and it shouldn't slope downwards at any point either, as that gives the ridiculous scenario where utility declines as wealth rises.

EU theory can be applied to both financial wealth outcomes and countless other everyday choices too. As long as its key axioms are met the theory can give insight into the motivation behind decision-making.

3.2 Ellsberg Paradox

Despite the appeal of expected utility theory its required axioms often don't hold up in practice. One such example comes from Ellsberg, who devised a two stage preference experiment involving different colour balls. The first stage involved participants being presented with a single urn containing 90 balls. They're told that there are 30 red balls inside, and the remaining 60 are either black or yellow in unknown amounts. The balls are well mixed with as much chance of picking one colour ball as another, and the drawing of balls is made without looking.

The participants are told that they have a choice of two gambles:

Gamble A: win £100 if a red ball is found on the first pick;
Gamble B: win £100 if a black ball is found on the first pick.

The participants in Ellsberg test showed a preference for gamble A over gamble B, and chose 30/90 or one in three odds of the red ball prize over unknown odds of finding a black ball, but which should have been assumed to also be 30/90 with no further information suggesting otherwise. This shows risk aversion to confirm the flaws of expected value theory and support EU theory, but that theory also fell apart when Ellsberg moved to the second stage of the test.

In the second phase of Ellsberg's experiment he used the same participants, urn, and the same 90 balls as before, with 30 red and unknown amounts of black and yellow balls making up the remaining 60. But this time two new gambles were offered:

Gamble C: win £100 if a red or yellow ball is found on the first pick;
Gamble D: win £100 if a black or yellow ball is found on the first pick.

This time the participants in the test showed a preference for gamble D and the 60/90 odds of winning the £100 based on the information they'd be given, over gamble C where they didn't know the odds beyond knowing that there were 30 red balls in the urn. This appears to be a confirmation of the risk aversion displayed in stage one of the test, with participants preferring to deal with known odds over unknown probabilities.

But the only difference between the first pair of gambles and the second pair is the addition of yellow balls to the gambles. The urn and the 90 balls inside are constants throughout, as are the presence of 30 red balls with the remainder being black or yellow, and each colour ball having the same odds of being drawn without looking.

Gamble A gave the £100 prize for finding a red ball on the first pick, and gamble C was more generous and added yellow to the choice. Gamble B gave the prize for a black ball, and gamble D added yellow to that gamble. With the (unknown) number of yellow balls not changing, it is a contradiction in preferences to prefer gamble A over B, and then gamble D over C, as the participants did. These preferences go against the required independence axiom of expected utility theory, and the model breaks down.

3.3 Allais Paradox

Further evidence against the assumptions of EU theory comes from experiments conducted by Maurice Allais, who also presented test participants with a series of gambles. In his first experiment participants were offered a choice of gamble 1A with a guaranteed 100% chance of £1 million, or alternatively gamble 1B with an 89% probability of £1 million, 10% chance of £5 million, and a 1% risk of getting nothing. Although expected value theory would suggest that gamble 1B is the better choice, the actual result of participants preferring gamble 1A with a certain return is in itself not a violation of expected utility theory.

Allais' second experiment saw him test the same individuals as earlier, who faced a choice between either gamble 2A, with an 89% chance of getting nothing, and an 11% probability of gaining £1 million, or alternatively gamble 2B with a 90% risk of nothing and 10% chance of gaining £5 million. This time the test participants preferred gamble 2B over 2A, which violates the independence axiom of expected utility theory, as the only difference between the number 1 and 2 gambles was the presentation of an additional option. The next diagram makes things clearer by simply rewriting the information about the gambles above in comparable terms. The 100% probability of winning £1 million in gamble 1A is rewritten as both 89% and 11% chances separately, and the 90% risk of winning nothing in gamble 2B is rewritten as both 89% and 1% chances separately too.

EXPERIMENT 1				EXPERIMENT 2			
Gamble 1A		Gamble 1B		Gamble 2A		Gamble 2B	
Prize	**Odds**	**Prize**	**Odds**	**Prize**	**Odds**	**Prize**	**Odds**
£1m	89%	£1m	89%	0	89%	0	89%
£1m	11%	0	1%	£1m	11%	0	1%
		£5m	10%			£5m	10%

As the diagram shows, in experiment 1 with gambles 1A and 1B the 89% chance of £1 million option is identical, and it may as well be removed for the purposes of deciding between the two gambles as it won't be a factor. In experiment 2 the 89% probability of winning nothing is exactly the same in both gamble 2A and 2B, and it also may as well be removed for comparison of the two gambles, as it can't possibly be a deciding factor.

EXPERIMENT 1				EXPERIMENT 2			
Gamble 1A		Gamble 1B		Gamble 2A		Gamble 2B	
Prize	**Odds**	**Prize**	**Odds**	**Prize**	**Odds**	**Prize**	**Odds**
£1m	11%	0	1%	£1m	11%	0	1%
		£5m	10%			£5m	10%

With the duplicate information removed from the gambles in experiment 1, while the other repeated information is cleared from the gambles in experiment 2, the situation is made far clearer. Gambles 1A and 2A are identical, as are 1B and 2B. The participants in Allais' study clearly violated the required independence axiom of EU theory by preferring 1A to 1B, and then 2B to 2A.

3.4 Preference Reversal Phenomenon

A further challenge to EU theory comes from what is known as the preference reversal phenomenon. A series of experiments have been conducted with participants presented two different bets. The first bet could be known as the P-bet, as it offered a higher probability (P) of a significant monetary gain, while the second bet may be called the £-bet, as although the odds of winning it were lower the pounds sterling (£) prize on offer was larger than in the P-bet.

The participants in the experiments were asked for both their preference of the two bet types, and to value each using methods such as their maximum bid to buy it or their minimum selling price if they owned it. While the test subjects have frequently shown a strong preference for the P-bet and its high probability for a significant monetary gain, when it comes assigning a monetary value to each bet it's the £-bet that gets the higher valuation, and they demand a higher selling price to sell that bet. This represents contradictory preferences to further disprove EU theory, as the participants' behaviour essentially says that while the £-bet is worth more money they prefer the P-bet, and that they prefer less money to more.

Preference reversals, the Allais paradox and the Ellsberg paradox suggest that people may lack the rationality to perform expected utility maximization as EU theory asserts. They may want to achieve the best option available but are constrained by their own limitations and biases. It may be more accurate to say that they are instead boundedly rational, and they don't seek to maximize utility but engage in satisficing, and try to get the most satisfaction possible.

4 Prospect Theory

4.1 Original Prospect Theory

As expected utility theory is based on assertions about human behaviour that are often missing in the real world, an alternative tool is required to model choice under risk and predict decision-making. Kahneman and Tversky's Prospect Theory is one such model, which incorporates the risk aversion or loss aversion common in individuals to challenge expected value theory, and also the need for a reference point evident with the preference reversal phenomenon that challenges expected utility theory.

Prospect theory shares common ground with expected utility (EU) theory in some aspects, and its subjective value (i.e. utility) function is found by testing tests participants' own valuation of lottery outcomes. For example, how much would they pay for a 50/50 fair coin toss to win either £10 or £0, and what about £20 or £0? But the key difference is that prospect theory doesn't create only one EU function with the data it collects, but two, with one for gains and one for losses. The data on the prices participants are willing to pay to gain each gamble is recorded as before, and then participants are told to imagine they owned the gambles, and are asked or tested on the lowest price they'd be wiling to accept to sell it.

The willingness to pay (WTP) to gain the gamble is thought to display risk aversion.

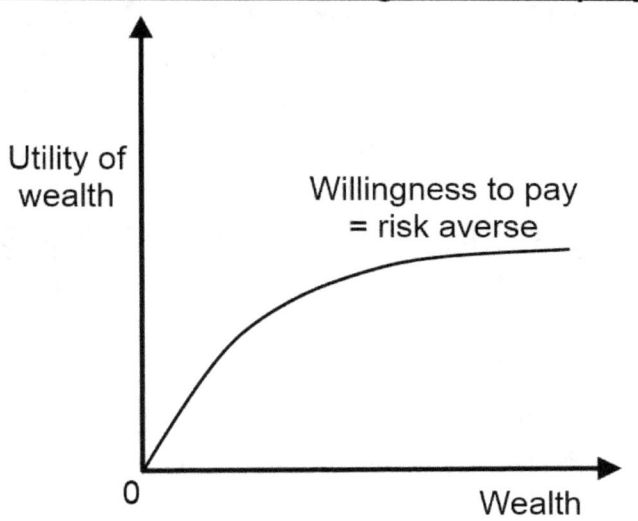

As the wealth (expected value) of the risky gamble rises, the utility gained grows at a slower rate, and as a result individuals are willing to pay (WTP) less to gain the gambles and the potential prizes on offer. For example, in 50/50 gambles to gain £10 or £0, £30 or £0, and £50 or £0, the pattern may go as follows:

£10 or £0 50/50 gamble = EV of £5, WTP of £4;

£30 or £0 50/50 gamble = EV of £15, WTP of £10.5;
£50 or £0 50/50 gamble = EV of £25, WTP of £12.5.

While the EV or wealth increases the willingness to pay barely moves, as individuals gain little from higher wealth and see no reason to risk money for a shot at it.

The willingness to accept (WTA) to sell and lose the gamble, once the participants in the test have been told to imagine that they've been given it, is thought to display risk loving or risk seeking behaviour.

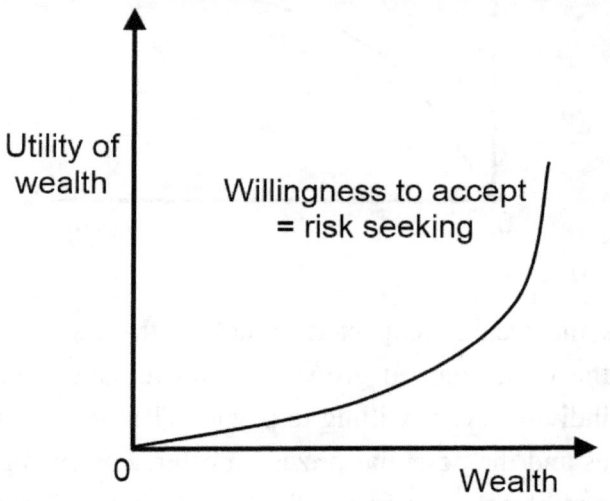

As the wealth (expected value) of the risky gamble rises the utility gained grows at a faster rate, the exact opposite of before, and individuals will demand a higher willing to accept (WTA) price to sell the gambles and potential prizes. For example, in 50/50 gambles to gain £10 or £0, £30 or £0, and £50 or £0, the pattern may be:

£10 or £0 50/50 gamble = EV of £5, WTA of £5;
£30 or £0 50/50 gamble = EV of £15, WTA of £18;
£50 or £0 50/50 gamble = EV of £25, WTA of £40.

While the EV or wealth increases the willingness to accept price demanded increases further, as individuals benefit less from a large gain in wealth than they suffer with an equivalent size loss. The WTP was kept down when the gamble was something to gain, as people were more worried about paying high for the gamble and losing the money they had if the prize ended up as the potential £0 option, than missing out on a huge gain they didn't yet have if the gamble came good. The WTA is kept high for the same reason, and people who own the gamble are more concerned about losing the (potential) value from the product they have, than gaining a certain payoff they don't yet have with a lower WTA which would be sure to attract buyers. It all appears to come back to loss aversion.

The value function of prospect theory combines the preference curves for gains and losses, to show how individuals value and attribute utility to a risky prospect under conditions of uncertain outcomes.

Prospect theory's value function

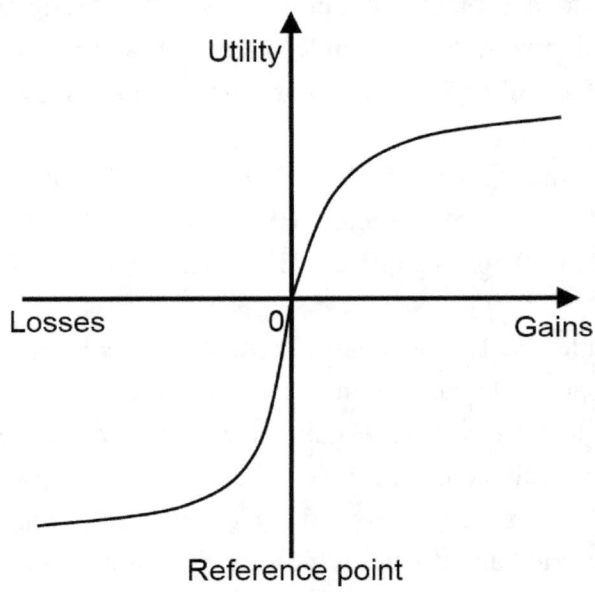

The value function sees a steep rising curve for losses, with risk loving when faced with the certain prospect of losing money as people try to escape this situation. For gains the value function sees a flattening curve, showing risk aversion when faced with the certainty of gaining money as people don't want to lose what they have.

Expected value theory evaluated value outcomes, expected utility theory looked at the utility gained from each value outcome, and prospect theory calculates the value prospects whether gain or loss. It is essentially a corrected utility curve that addresses the flaws of expected utility theory, where gains and losses were not acknowledged as having different qualities.

But prospect theory does make a major change to expected value and expected utility theories. While other theories weigh the value outcome against its actual probability of occurrence, with prospect theory this weight isn't based on a probability of occurrence, but is instead a decision weight that acts subjectively on that probability. The probability of a loss or gain may be 1% but it will not necessarily be assessed and acted upon as 1% in prospect theory, but instead as how human biases and limited rationality predict humans would treat it. Remember that the point of a theory to model choice under conditions of risk is not to show what should happen, but what does happen in reality. The decision weights used by prospect theory are designed to mimic real human behaviour.

Prospect theory's decision weights on probabilities, $d(p)$, share some features with actual probabilities, p, with d an increasing function of p, and the higher the probability the more likely that outcome will be the one decided upon, as would be expected. Very low probabilities receive overweighting, with decision weights that ensure $d(p) > p$, and combined with the separate overestimation of low probabilities this ensures rare events have a greater impact than they should. This appears to be an inherent characteristic of human nature, and people worry more about very rare events such as being involved in a plane crash than the more common risks of daily life.

While very rare probabilities are overweighted, this is overcome by a general trend for decision-makers to underweight the actual probabilities of outcomes coming to pass, and the decision weight shows a greater divergence from the actual probability when probabilities are high then when they are low. For example, smokers feel that they can light up and suffer no consequences despite the well publicized risks, while daredevils are sure that although others may suffer injury by performing their stunts, they can use their natural skill to walk away unharmed. This tendency for people to underweight greater percentages can explain why product prices often end with .99 or .95, and sellers know that buyers are likely to see such a price as lower than it actually is.

The result of this general probability underweighting sees the sum of all decision weights total a lower amount than the sum of all probabilities. Although all probabilities will sum to one, decision weights will total less than this:

$$d(p) + d(1-p) < 1$$

A change in probabilities will not always be met with the change in preferences that would be expected, and people may not adjust properly to changing circumstances. This means prospect theory's decision weighting function isn't well behaved as it nears the end points and clear preferences associated with certain outcomes. For this reason probabilities very close to 0 or 1 are rounded to these respectively, and the corresponding d(p) = p, will see d(0) = 0, and d(1) = 1, where the probability will be acted upon without a need for a decision weight.

Prospect theory's decision weighting function of probabilities can be seen with the dashed line, where the 45 degree line is what the decision weights, d(p), would be if they acted directly on probabilities, p. These weights are then applied to the value outcomes for gains or losses.

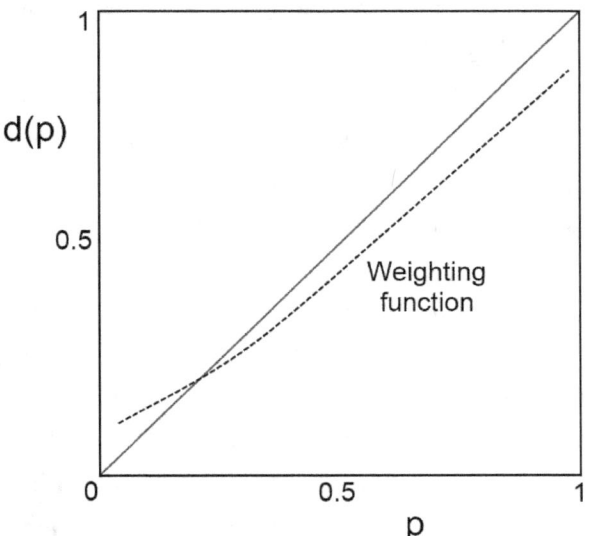

4.2 Cumulative Prospect Theory

Despite the improvements prospect theory offers on expected utility (EU) theory, with the inclusion of a reference point and awareness of the loss aversion common in individuals, it still exhibits many flaws. While it may show awareness of the way human preferences change between losses and gains, it still can't predict preference reversals, one of the main flaws with expected utility theory. This stochastic dominance, the random selection of previously dominated and less preferred options as the new dominant preference, can't be allowed for the model to function properly as a predictor of human decision-making. The other major problems are the poorly behaved probability weightings near reference points of certain outcomes, and the sum of decision weights not summing to one as probabilities do.

Cumulative Prospect Theory (CPT) builds on the earlier model and directly targets the source of its flaws, which is the lack of distinction made between gains and losses in the decision weighting function. Original prospect theory (OPT) assigns each probability a unique decision weight and applies this equally to gains and losses, and while it acknowledges that gains and losses are valued differently it ignores the possibility of each having different probability decision weights too. This can explain the preference reversal phenomenon seen between gains and losses, and also the problems found with uncertain probability weightings at the reference point margins where gains become losses, as seen in the image at the end of the last section.

In CPT each probability, p, doesn't have its own decision weighting, d(p), to be applied equally to positive gains or negative losses. It instead shares a cumulative weighting function, as the sum of all of decision weightings is made to total one:

$$\text{OPT: } d_-(p) = d_+(p)$$
$$\text{CPT: } d_-(p) = 1 - d_+(1-p)$$

Experiments involving test participants were used to develop CPT into a model that could be used to predict choice under conditions of risk, and the results found a fourfold pattern of risk attitudes. Earlier research had found only risk aversion for gains and risk seeking for losses but CPT went into greater detail. It predicted risk aversion for moderate or high probability gains and risk seeking for moderate or high probability losses, while low probability gains saw risk seeking and low probability losses saw risk aversion. This may be explained by the overweighting of low probabilities for both positive and negative prospects, and the underweighting of moderate or high probabilities.

Decision weightings for gains, d(p)+, and losses, d(p)-, are shown in the graph. The thicker dashed curve representing decision weightings for gains is more curved than that for losses, suggesting that people have slightly stronger opinions when gains are on offer than when faced with losses. This cumulative weighting function overcomes many of the problems seen earlier with original prospect theory, and most noticeably it is well behaved around the reference points of 0 and 1 probabilities, instead of being random and problematic.

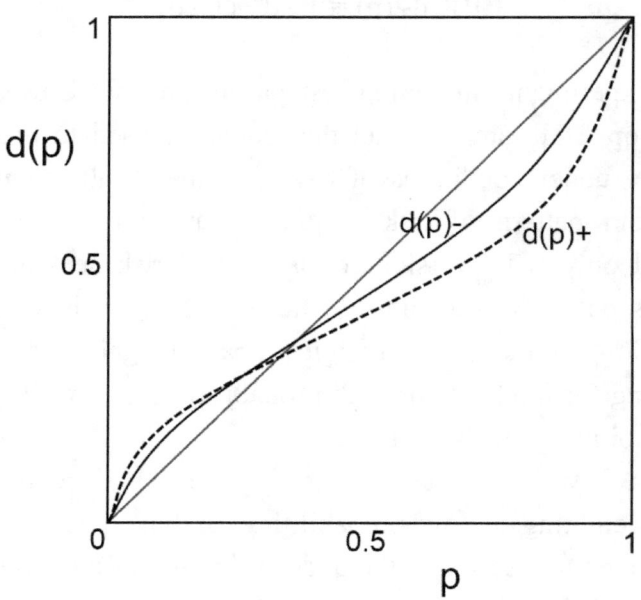

Cumulative weighting function

The changes made in CPT resolve the problems around decision weightings that hindered the original model, replacing set weights for each probability with a cumulative weighting function that incorporates gains and losses. But the most serious problem with original prospect theory (and expected utility theory) remains, as this amended model also can't predict preference reversals. Without this ability the model won't represent what actually goes on with human behaviour, and it will have little use as a tool to predict future behaviour. Further developments are needed to tackle the preference reversal phenomenon.

4.3 Third Generation Prospect Theory

Two different types of lottery are commonly used in test experiments to find peoples' preferences; a P-bet lottery with a higher probability of a payout but lower prize in monetary terms, and a £-bet lottery with a lower probability of a prize but a higher payout in pounds sterling if won. The preference reversal phenomenon gives the strange situation where the P-bet is preferred in a straight choice over the £-bet, but the £-bet is given the higher valuation. The problem is thought to come down to the £-bet being given too high a valuation, when test participants are told to imagine they own the gamble and have to suggest their lowest selling price.

Loss aversion is the likely cause of preference reversals, and when faced with gains (such as giving a preference or bidding) individuals are known to display risk aversion, which pushes them toward the safer P-bet, yet when faced with losses (such as selling the lottery) they display risk seeking behaviour to avoid the loss, which pushed them to overvalue the more lucrative £-bet when selling it.

Previous versions of prospect theory have tried to model and incorporate the loss aversion intrinsic to human beings by including a reference point, separating gains and losses instead of focusing only on value outcomes. But the reference points used have looked at certain outcomes, with 0 representing a certain loss and no gain at all while 1 represents a guaranteed gain, and yet the £-bet at the heart of the preference reversal problem is not a certain outcome but an uncertain lottery. For the problem to be modelled and understood the reference point must have the same characteristics of the £-bet, where a person holds ownership of an uncertain lottery with the risk of winning nothing and a small chance to win a large monetary payout.

Third generation prospect theory (3PT), devised by Schmidt et al., includes uncertain reference points for the first time in the history of prospect theory. It does this by using the idea of multiple possible states, as gains or losses, decision weights, and preferences are now defined separately for each and every possible situation. With this increased uncertainty it can now represent the situation where participants are endowed with an uncertain lottery.

The parameterization of 3PT includes varying percentage odds of winning the higher probability P-bet, and lower probability £-bet, and a ratio of the two lotteries. It adds a parameter giving the level of indifference between gains and losses, denoted by L and showing the loss aversion, and another showing the curvature of the value function, denoted by S and showing the sensitivity to risk. With these parameters the model can be tested to see if it can predict preference reversals and accurately represent human behaviour.

Choice (preference) and valuation (selling price) functions were created based on parameters already proven valid, based on the results of experiments conducted in earlier versions of prospect theory, and then applied to the P-bet and £-bet separately. The parameters used were altered up and down to move between situations where the P-bet or £-bet received the higher value in each, and would be preferred in a choice or given the higher selling price respectively. The preference reversal phenomenon sees the P-bet preferred in a straight choice over the £-bet, yet the £-bet is given the higher 'willingness to accept' (WTA) selling price valuation, and this is the situation the model needs to predict.

The following diagram shows some of the results of the tests of 3PT. The sensitivity to risk parameter, S, moves from a range of 0.1 on the left to 1.9 on the right, and a value of 1 in the centre of the image represents indifference between the safer P-bet and the riskier but more lucrative £-bet. With S values below 1 there is diminishing sensitivity to risk, and the safer P-bet is more appealing and would be chosen over the £-bet, but with S values over 1 there is increasing risk sensitivity and the low probability but lucrative prize of the £-bet is preferred. The loss aversion parameter, L, moves from a range of 0 at the bottom of the image to 3 at the top, with a value of 1 showing loss neutrality. Values above 1 show loss aversion, and above this point the £-bet has the higher WTA selling price valuation, unless a low sensitivity to risk under 1 makes up for the loss aversion to give the P-bet the higher WTA price.

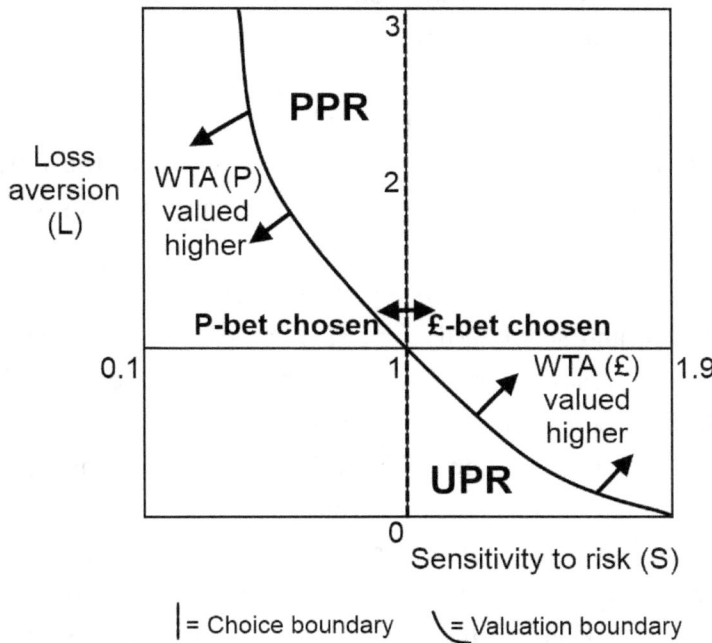

Preference reversals can be seen in this diagram, and it looks like the third generation version of prospect theory is finally able to predict them. The situation where the P-bet is preferred but the £-bet valued higher when being sold is visible in the area with the letters PPR, standing for predicted preference reversals. This will occur if the risk sensitivity or value function curvature has a value below 1, while the subject also exhibits some loss aversion with a level above 1. The P-bet being chosen is the easier criterion to achieve in a parameter test and only requires diminishing risk sensitivity, but for the £-bet to receive a higher selling price valuation this cannot be too low, or will need to be made up for with high levels of loss and risk aversion.

Interestingly the image also shows a new and unpredicted form of preference reversals, where the £-bet is chosen but the P-bet valued higher, and this is shown with the area labelled UPR, standing for unpredicted preference reversals. This again depends primarily on loss aversion and secondarily on risk sensitivity, but the value of loss aversion should be below one not above it, while risk sensitivity should be above one.

Beyond this image 3PT did find certain circumstances of parameter values where preference reversals occurred without any loss aversion, and all that was required was diminishing risk sensitivity. Although loss aversion and risk aversion is a driving factor throughout the tests, this exception shows that there don't need to be strong preferences for there to be preference reversals, and this may say a great deal about the uncertain nature of human behaviour and the difficulty in modelling it.

Third generation prospect theory includes uncertain reference points to finally predict the preference reversals that have blighted earlier versions, and the popular expected utility theory model. Although the model appears able to show the behaviour that humans exhibit during choice under risk, 3PT is not without its flaws. Original prospect theory took the original expected utility model and generalized it, replacing the fixed value function with the idea of prospects and ever changing context, before cumulative prospect theory generalized the weighting function to replace the previously fixed probability weightings. Third generation prospect theory continues this process, and generalizes reference points to allow uncertainty there too. While this may make a more accurate model in terms of representing individual human behaviour, the constant generalizing risks creating a theory that doesn't stand for anything, and at some point it may cease to function as a proper model and be of little use.

5 Risky Games

5.1 Certain and Uncertain Games

The previous sections have looked at choice under risk, where people navigate uncertain outcomes with known probabilities of occurrence, and their risk attitude determines their behaviour. This chapter adds uncertain probabilities to the discussion, as individuals no longer face outcomes alone but with others who can affect their result. The choices take place within the framework of the simultaneous two player interaction of game theory, but without the usual game theoretic assumption that participants follow the strategy that gives them the best payoff from the interaction (known as a game), given the behaviour of the other player (person interacting). Instead of only trying to maximize their expected utility as game theory typically follows, individuals will also act on their risk attitude given their prospects, whether risk seeking for losses or risk aversion for gains.

Although game theory is a very broad category this chapter will focus on a very specific type of game, known as a zero-sum game. This is where the interests of those involved in the game are diametrically opposed and where one player's gain is another's loss, giving the interaction payoffs a sum of zero and creating the zero-sum game. No other type of game includes both gains and losses as outcomes, and ensures that one of the two will come to pass. This is important as the focus is the human reaction to choice under uncertainty, and the most uncertain situation is where an individual knows that their choice can decide whether they face a gain or a loss, or at the very least the level they will face.

Game theory often involves non-zero sum games, where players can either both gain together or both lose together, and this situation is of limited relevance here as a player can avoid having to face uncertainty. In games like these there is often a dominant strategy for a player, where an individual's incentives will push them to follow a certain course of action no matter what the other player does.

The next image shows a typical game, where player one's options are in the rows across and player two's in the columns' down, and they can either cooperate or fight each other in a battle for power. The numbers in the matrix are known as payoffs and the first or left payoff in each section is for player one, while the second or right payoff is for player two. It's assumed that players prefer larger payoffs, represented by greater numbers in the grid.

In the event that player two cooperates fighting would give player one a payoff of 3 instead of 2, and if player two fights then doing the same would give player one a payoff of 1 in place of the -1 payoff linked with cooperation. Player one's dominant strategy is to fight and that will always give the best individual outcome, and with symmetrical payoffs for both players the same holds for player two. The result of this game is predictable and both will fight, ensuring that there's no need for an uncertain choice to be made.

A dominant strategy

	Player 2 Cooperate	Player 2 Fight
Player 1 Cooperate	2, 2	-1, 3
Player 1 Fight	3, -1	<u>1, 1</u>

The next game has a different format. It is a chicken game where two drivers face their cars at each other in a duel and wait for the other to break first. The payoffs are identical for both players but there isn't a dominant strategy here, as both the swerve and the keep the car straight strategies can outperform the other depending on the actions of the rival player. If the other player swerves then keeping the car straight is the best option, for a payoff of 1 through the act of one-upmanship instead of swerving for a 0 payoff. And if the other player drives straight at you then to swerve is best, and even though it gives a -1 payoff through acting as a coward it beats the -10 payoff from a crash.

A chicken game

		Player 2	
		Swerve	Straight
Player 1	Swerve	0, 0	-1, 1
	Straight	1, -1	-10, -10

But while the swerve/straight or straight/swerve outcomes are zero-sum and see one player gain what the other player loses, an uncertain choice doesn't necessarily have to be made here. Player two's payoffs are exactly the same as player one's for each course of action, and if player one could make an agreement with player two then both could swerve their cars and be sure to avoid a loss, securing the 0 payoff and ending up no worse than when they started. Just like the game above this interaction is not necessarily about individual choice under uncertainty, but could come down to cooperation to avoid mutual losses.

Even without cooperation there is still a long-run outcome here, the Nash equilibrium where neither player can improve their payoff without the other player changing their course of action. It's underlined here and occurs where one player swerves and the other stays straight. At this point the swerving player would move from a -1 payoff to a worse -10 if he returns to the straight path, while the straight driver would move from a 1 payoff to 0 if he swerved. Once the players reach this point their incentives should keep them there and there's no need for a choice to be made.

A zero-sum game differs from those here by including uncertainty at all levels of the simultaneous two player interaction. It sees an uncertain strategy for each player and not a clear individual dominant strategy, uncertain outcomes in place of a clear long-run Nash equilibrium, while a player must also face the uncertainty of acting alone as there's no possibility for cooperation with payoffs diametrically opposed.

5.2 Minimax Strategies

The game below shows a simultaneous zero-sum game between two players; player one and player two. Player one's options are to either choose strategy 1A or 1B while player two's are to select either 2A or 2B, and as the diagram below shows the payoff for one player is the exact opposite of the payoff for the other. This means a player can only achieve a gain by forcing an equivalent loss on his rival, and he will always lose if the other player wins.

A zero-sum game

		Player 2	
		2A	2B
Player 1	1A	-4, 4	3, -3
	1B	5, -5	-2, 2

A glance at the four payoff numbers for player one reveal that he doesn't have a dominant strategy here. Choosing 1A could either get him a payoff loss of -4 or a gain of 3 depending on player two's action, while choosing 1B might see him get a payoff gain of 5 or cause him to suffer a loss of -2.

There's also no hope of cooperation here as unlike in other games it can never be mutually beneficial, because of the zero-sum adversarial nature of the game. Anything that benefits one player hurts the other player to the exact same extent, and both players know it. This feature also prevents a long-run Nash equilibrium, as whatever the outcome one of the players will be better off changing their choice of action, and the best outcome for player one is the worst for player two, the second best for player one is the second worst for player two and so on.

To deal with the uncertainty of a zero-sum game players may use a number of risk covering strategies, to prepare for either a good or a bad outcome:

Minimax refers to minimizing the maximum loss, and involves a player first assuming the worst case scenario where a loss is forced upon them and then selecting the strategy that is linked with the highest available payoff, to make the best of the losing outcomes and avoid a high loss. It could be seen as exhibiting risk seeking behaviour for losses. In a zero-sum adversarial game minimax also minimizes the opponent's maximum gain;

Maximin refers to maximizing the minimum gain, and involves looking at the best case situation where they secure a gain, and then selecting the strategy linked with the highest minimum payoff, to ensure they will at least achieve a certain gain. It could be seen as exhibiting risk aversion for gains. In a zero-sum adversarial game maximin also maximizes the opponent's minimum loss.

If player one applied a minimax strategy to this game then he would assume the worst and foresee a loss, which may be either the -4 or -2 payoffs (payoff gains of 4 and 2 for player two respectively). To minimize his maximum loss he'd go for the strategy linked to the -2 payoff.

If player one instead applied a maximin strategy to the game, assuming the best and that he would gain a positive payoff, the possible outcomes for him would be a payoff of 3 or 5 (payoff losses of -3 or -5 for player two respectively). To maximize his minimum gain he would go for the choice of action linked to the payoff of 5.

Player one's minimax strategy to chase a payoff of -2 and his maximin strategy to go for a payoff of 5 are both linked to the same choice of action. They are the two possible outcomes for player one if he selects strategy 1B:

	2A	2B
1B	5	-2

It is therefore safe to assume that player one will choose B out of his two options, and the only question is what player two will do. His payoffs are the exact opposite of player one's, and if player two followed a minimax strategy of his own then he would assume a loss, which could be -3 or -5, and then choose the action linked to the payoff of -3. If he instead followed a maximin strategy then player two would assume a gain which could be 2 or 4, and he'd go for the action linked to the 4 payoff. The minimax strategy pushes him to go for 2B, but the maximin strategy pushes him to go for the opposite and 2A. Player two would therefore have to decide between minimax and maximin strategies.

To keep things simple the minimax strategy will be chosen for both players. This is more defensive than maximin and more relevant to an adversarial zero-sum game, and when you know that your opponent can hurt you the first impulse is to defend against possible losses.

With both players going for minimax the results of the game can be predicted. Player one will choose 1B for the reduced loss of -2 (reduced payoff gain of 2 for player two) if a loss were forced upon him, while player two will choose 2B too for the reduced loss of -3 from his point of view (and a reduced gain of 3 for player one). That gives an outcome of 1B, 2B where player one suffers a -2 loss and player two gets a 2 gain.

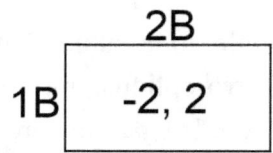

But the 1B, 2B outcome won't last long, and knowing that player two is at 2B player one will switch his choice to 1A to get the 1A, 2B payoff gain of 3 for himself and a -3 loss for his rival. This outcome is also unstable as player two will realize he's suffering and change his behaviour to follow 2A, securing the 1A, 2A outcome of a 4 payoff and forcing a -4 loss on player one. But then player one switches back to 1B to avoid that loss, and the result is the 1B, 2A payoffs of 5 for player one and -5 for player two. The final step is for player two to return to a B strategy, for 1B, 2B payoffs and the same outcome as originally. Then the cycle begins all over again. The diagram below shows how players move between the short-run outcomes.

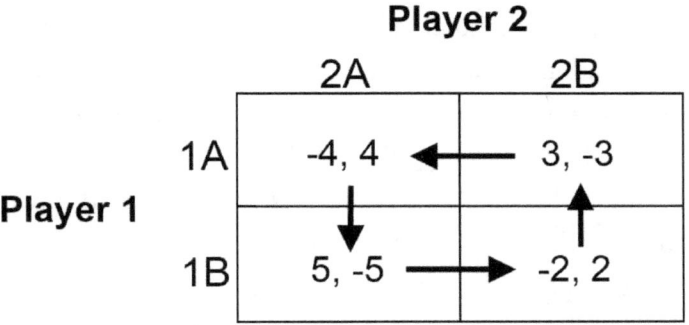

To find a stable outcome to this game each player wouldn't be able to use only one strategy but would instead have to adopt a mixed strategy. If they didn't then they would have a difficult time making a choice of action, and trying to stay ahead of the ever changing movements of the other player.

If player one were to play 1A 50% of the time and 1B 50 % of the time then, in the event that player two chose to play 2A, the expected payoff for player one would be:

$$- 4*(0.5) + 5*(0.5) = - 2 + 2.5 = \underline{0.5}$$

And if player two were to choose to play 2B the payoff for player one is:

$$3*(0.5) - 2*(0.5) = 1.5 - 1 = \underline{0.5}$$

Player one's mixed minimax strategy minimizes the expected loss and actually offers a gain of **0.5**. This is far superior to the -2 expected payoff he was hoping for when he first adopted a (non-mixed) minimax strategy that played 1B all of the time to minimize his maximum loss. It's also a stable strategy that can't be ruined by player two and will hold irrespective of his behaviour. It's worth noting that the 0.5 outcome is simply the average of player one's four possible payoffs (-2, -4, 3, 5), but the mixed minimax strategy removes the influence of his risk attitude which could have led him to an inferior payoff.

There is also a mixed minimax strategy that will get player two a better minimax payoff, found with trial and error of different percentage weightings of 2A and 2B in an Excel spreadsheet. If player two plays 2A 35.7% percent of the time, and 2B 64.3% of the time, then in the event that player one chooses 1A then player two's payoff will be:

$$4*(0.357) - 3*(0.643) = 1.428 - 1.929 = \underline{-0.501}$$

And if player one were to play 1B the expected payoff for player two is:

$$-5*(0.357) + 2*(0.643) = -1.785 + 1.286 = \underline{-0.499}$$

The result is a minimax payoff of **-0.5** for player two, far better than the -3 payoff player two was hoping to get by playing a (non-mixed) minimax strategy of choosing 2B all of the time to minimize his maximum loss. And the strategy will hold irrespective of player one's actions, so there's no need to change strategy constantly to try and stay ahead. As with player one the -0.5 outcome is simply the average of player two's possible payoffs (-3, -5, 2, 4).

By adopting a mixed minimax strategy both players can improve their worst case scenario payoffs and achieve a stable strategy that can't be improved upon, and the mixed minimax strategy that each player follows becomes the long-run Nash equilibrium of the game.

6 Auction Theory

6.1 Types of Auction

Auctions are one of the most interesting areas in choice theory, as unlike other fields the outcome isn't necessarily uncertain, and an individual holds the power to get the outcome and prize they want. The uncertainty comes from not knowing how to achieve this, as bidders are unsure how to secure their desired item while not wasting their money needlessly. They don't know if there are countless others forcing up the bid price required to win, or if they're the only ones going for it and a lower bid price will both win the auction and save them money.

There are various different formats of auction, and each type has its own unique traits to note:

Common Value Auctions are where the auction item's value is the same to all bidders, but each bidder has different information about its true value. An example of this would be a land or property. Every bidder would value the land or property the same as it has a certain value, but they may not know for sure what that value is due to varying levels of knowledge about the wider market;

Private Value Auctions are where each bidder values the item differently, and their valuation is made privately and completely independently of other bidders. The values differ because they are based on the bidder's desire for the item, which will vary by individual. This is the most common type of auction and an example is second hand items sold over the internet on eBay. Private value auctions can be divided into three types:

1) **English Auction**, also known as an ascending bid auction, sees a price that starts low and is gradually raised until only one bidder remains. This is probably the most familiar and popular type of auction and can be seen at any auction house, where an auctioneer sits in front of a room of bidders and watches out for gestures or shouts;

2) **Dutch Auction**, also known as a descending bid auction, sees a price that starts artificially high and is lowered until someone is willing to buy and pay that amount;

3) **Sealed Bid Auction** is where bids are effectively given in sealed envelopes, and unlike other auctions an individual can make a bid without others knowing what it is until the auction is finished. Sealed bid auctions can be further divided into two different variations:

A) **First-Price Sealed Bid**, where the winning bidder pays the exact amount that they bid;

B) **Second-Price Sealed Bid**, where the winning bidder only pays the amount equal to the second highest bid, and the highest amount offered by other bidders. This is a more forgiving type of auction, and it reassures bidders that there is no risk of paying far more than the other bidders were offering, if an individual had bid far more than was required to win the auction.

6.2 Private Value Auctions and Revenue Equivalence

Bidders may wonder how their optimum bidding strategy will change with the different types of auction, and how they can avoid paying more than they have to. Imagine that there are 10 bidders competing to win the item up for auction. A bidder called Victor values the item considerably higher than anyone else, and he's willing to pay as much as £200 for it, but of course he would prefer to pay less than his maximum price. With this information each of the auction types can be run through in turn to find their individual results and winning bid price, with the assumption that bids can be made in increments as low as £0.01, and that the second highest bidder will offer £180 and not a penny more.

In an English auction the result is very simple. The bidding will start low and the ten bidders will watch what the competition does and slowly bid up the item price, until it exceeds their own individual private valuation. The second highest bidder will bid £180, and then Victor will offer £180.01 and claim the item.

A second-price sealed bid auction will have a very similar result to the English auction. Victor will bid his highest valuation of £200, and although the bids are sealed and hidden he has good reason to be honest about it. The price he bids doesn't determine the amount he pays in a second-price auction, only whether he wins. Bidding less than his true valuation risks losing the item to someone who values it less than he does, and bidding more means he may win but only because he outbid someone who values the item more, and he's paying more than he feels it's worth. Having bid his honest valuation of £200 Victor will be relieved when the hidden bids are examined, as he finds that he only has to pay the second largest bid of £180.

Winning bids in the English and second-price sealed bid auctions here are essentially exactly the same, £180 ≈ £180.01, and Victor doesn't need to worry about wasting his money, while the auction seller doesn't need to worry about choosing the wrong auction type. The revenue equivalence theorem states that the revenue going to the seller, and as a result the winning bid amount, will be equivalent in a private auction across all auction types. The two remaining types of private value auction can be examined to prove that the seller's revenue/winning bid will be the same.

In a first-price sealed bid auction Victor would have to assume that his valuation is the highest, and if it wasn't then he wouldn't want to win anyway as it's too costly as noted earlier. But he shouldn't bid £200 based on his valuation of the item, and he should instead make an offer of what he believes will be the second highest bid, to just edge it out and save money. The situation is similar to a second-price sealed bid auction, except that Victor isn't automatically given the second priced bid amount and he needs to find it for himself.

Because this is a private value auction bidders decide their values independently, and observe no other information about other bidders' value before they bid. When Victor is trying to guess the second highest bid to decide his optimum bidding price he'll have to make some assumptions, and with no evidence to the contrary he'll assume that the bids of the 10 bidders (including him) are uniformly distributed, from the lowest to the highest. If valuations are uniform then the second highest bid (SHB) can be found from the highest bid (HB) and the number of bidders (N):

$$SHB = (N - 1)/N * HB$$
$$SHB = 9/10 * 200$$
$$SHB = \underline{£180}$$

As Victor has figured out that the second highest bid will be £180, he can make a bid just above this at £180.01 and win the first-price sealed bid auction without having to spend more money. This is the same price as earlier. A Dutch auction would follow a similar pattern and Victor would know that the second-highest bidder would offer £180, and he could accept the descending auction price and win the auction at a level just above this at £180.01, an equivalent level to support the revenue equivalence theorem.

The general result that revenue equivalence holds in private value auctions irrespective of auction type is encouraging, and an object's price shouldn't depend on how it is bought, but only on its innate value and the value competing bidders attribute to it. But this is of course only a general result, and the risk attitude of the bidders or seller could make all the difference in reality. For example a seller could receive greater revenue by selling in an ascending price English auction than a descending price Dutch auction, if bidders had high levels of loss aversion. In the English auction a bidder who has the high bid, outbidding the competition as the price slowly rises up from a very low level, may feel that he 'has' the item. From that position a bidder's loss aversion could come to the fore, and he would be bidding to hold on to the item not to gain it, and that valuation may be entirely different from the one he'd offer to gain an item in a Dutch auction format.

6.3 Common Value Auctions and the Winner's Curse

Common value auctions differ from private value auctions as bidders don't have their own unique valuations, but different information over the true valuation of the object. Bidders will take an interest in how others value their target, knowing that other people may have information that they don't. Because of this revenue equivalence may not hold as valuations depend on what others will pay, and a bidder may believe others will pay more or less depending on the type of auction format. For example, seeing the price descend continuously in a Dutch auction without anyone stopping it will reveal nothing, while watching bids rise in an English auction could reveal more about what others think and push up a bidder's individual valuation.

In a common value auction a winning bid is based on the highest estimate of the object's true value, and this puts bidders at risk of the 'winner's curse' where they overpay, as their expectation of the value was above average and their bid was the most upward biased. To avoid suffering the winners' curse bidders need to shade their bids, below the level indicated by their information. That involves figuring out the true value of the object in the event they were to win the auction, when they would have overestimated the true value relative to others. A bidder needs to use his information on a winning bid to figure out the average estimation of the object's real value, to then adjust his own estimation to this level.

A bidder would assume that his estimation of the object's true value is the highest estimation (HE), and if it isn't then he wouldn't want the object anyway as he'd have to pay more than he thinks it is worth. Using the number of bidders (N) and assuming a uniform distribution, the highest average estimation (HE) is:

$$HE = U * (N - 1)/N$$

The U represents the upper limit of the distribution, and solving for U gives:

$$U = [N/(N - 1)] * HE$$

The average estimate (AE) is what a bidder is interested in to bid no higher, and that would lie halfway through the distribution and could be found by dividing by 2:

$$AE = [N/2(N - 1)] * HE$$

As the formula shows, this average estimation is clearly going to be far lower than the highest estimation level that a bidder would have bid, and therefore it will help minimize the risk of the winner's curse.

For example, if there are 100 bidders (N = 100) who are considering making a bid on the property in the common value auction, and an individual bidder estimates the value at £200,000 (HE = 200,000), then the upper limit of the distribution is:

$$U = 100/(100 - 1) * 200,000$$
$$U = £202,020$$

This is the upper limit, and the lower limit will be £0. To find the average estimation (AE) the upper limit simply needs to be divided by 2.

$$AE = \underline{£101,010}$$

So the bidder will not bid his own estimate of £202,020, which he believes to be the highest estimate of the object's true value and therefore an overestimate relative to others, and would see him suffer the winner's curse of paying too much. Instead he will bid £101,010, or just above, which is his best guess of the average estimate value.

As the number of bidders rises in an auction the likelihood of the winner's curse rises too, as more bidders mean more bids and more information suggesting that the object's real value is higher, and that an individual needs to raise his own estimation and bid in accordance with this. But the price-raising effect caused by more bidders may be offset by a price-shading effect, and as the equation above shows, more bidders will reduce the average estimation and the amount that an individual actually bids.

7 Voting

7.1 Voter Preferences and the Condorcet Winner

Choice under uncertainty is made increasingly uncertain when other large numbers of other people come into play. And in the area of democracy where numbers are the only thing that counts the biggest factor is always going to be other people. Voting is considered one of the most important rights in a democratic society, and a choice of candidates or options are presented to individuals before the collective popular opinion decides which gets the go-ahead. Every individual voter will be outnumbered, and both their choices and the collective choice can be uncertain as a result.

If voting is only on a single issue, and voters have a clear preference where they gain the highest utility from that option and lose utility the further they move away from it, known as 'single-peaked' preferences, then the median position will win under majority rule. Although it may not be the first choice for many voters they will prefer it to the alternatives. Consider the three numbers below and imagine them as alternative options put before voters, and the only ones that are available.

1
2
3

1 is one choice, 2 is close to it in detail but a little different, and 3 is quite similar to the second option but a long way from the first choice put before the general public. If voters had a clear preference between the three and the issues were all single issue, then the median option 2 would win every time. Choices 1, 2, and 3 would each have their own dedicated supporters, but median option 2 also benefits by being the second choice for all of those who would first vote for option 1, and also all of those who would prefer option 3. This greater support would tip the scales in favour of option 2.

The median voter theorem insists that the median voter's preferences will win out under majority rule, but while this may hold for single policy issues in practice the issues that will be voted upon are multidimensional. In this situation the median position will not always win and choices need to be calculated.

Consider three individuals, A, B, and C, who each put their opinion on a multi-dimensional proposal forward, and then put it to a vote between the three to see which is most popular. Person A may have ideal point A, with declining utility as he moves away from this point. The image below shows person A's ideal point A in two-dimensional space, with a certain position (the place on the vertical) for one half of the two-dimensional issue put before the three voters, and a certain position (the place on the horizontal) for the other part of the multidimensional issue voters face. The lines curved away from the point A are known as indifference curves, and at any point along a curve there is indifference, where person A would just as happily be at the left end of the curve as the middle or right.

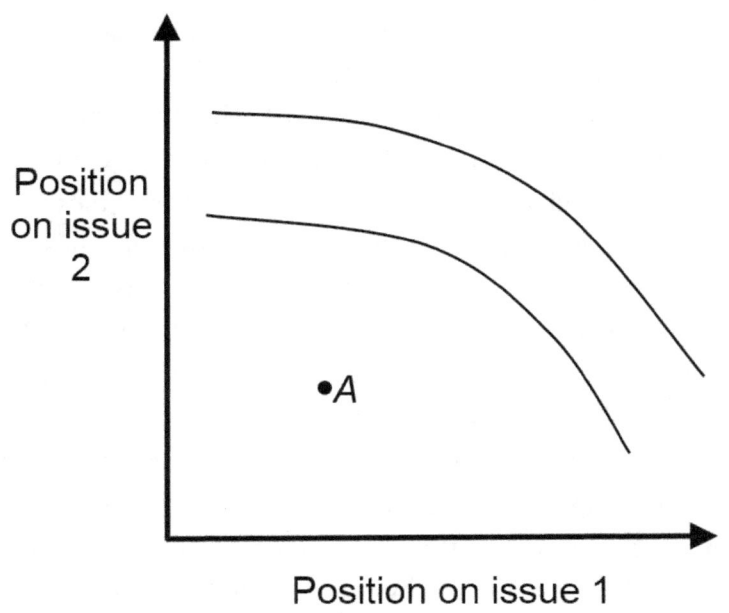

While point *A* is the ideal point for person A, any point on the curve closest to it is worse, but still better than at any alternative point along the indifference curve further away. As far as person A is concerned point *A* is best, the area within the first indifference curve out is second best, within the second indifference curve out third best, and finally the positions beyond the second indifference curve in the distance are the worst of the options. The area behind point *A* in the opposite direction to the indifference curves is considered 'out of bounds' and unwanted, where voter A doesn't even have a preference. He would never accept a position to the below left of his ideal point *A*, and his preferences run from point *A* as an ideal and then out toward his first and second indifferences curves and beyond as the less preferred options.

A second image adds person B to the discussion, and his ideal point is point *B*. Like person A his preferences are represented by the indifference curves, but these are dashed instead of solid to distinguish them from those of the other voters. Any point within the first dashed indifference curve is second best to point *B* for person B, a point within the second dashed indifference curve out is third best for him, and anything beyond the second dashed curve in the distance is least desired for him although still a possible option. The area behind proposal *B* and away from his dashed indifference curves to the up right is completely unacceptable by voter B, and he would never even consider a point here. His preferences run from point *B* as an ideal point toward his first and then second indifference curves and beyond only.

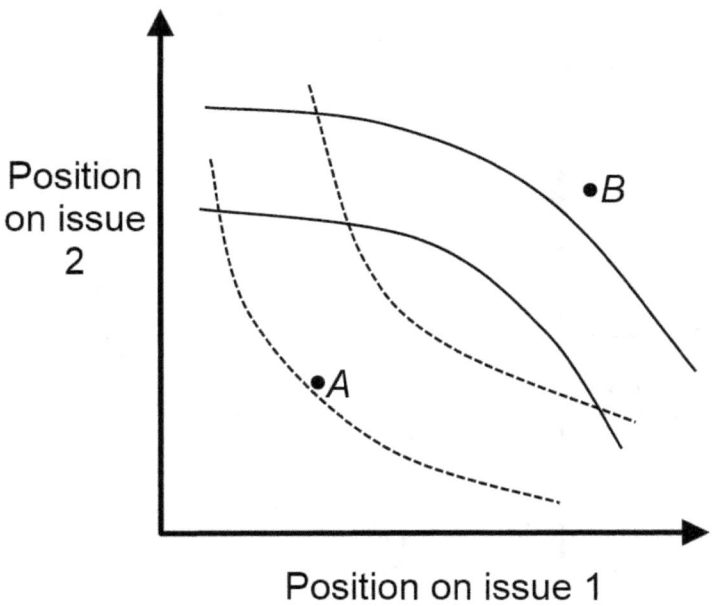

Position on issue 1

The final person to join the vote is person C, who predictably values point *C* as her ideal point. Her indifference curves are dotted to separate them from the others, while the fact that they are far closer together than the other peoples' is also a distinguishing factor. Like the two other people any point within the closer dotted indifference curve is her second best option, a place within the second dotted curve is next best, and anywhere beyond the final indifference curve is her worst option. In a similar way to the other voters, the area behind point *C* to the below right and away from the indifference curves is unacceptable to voter C, and she would only go for point *C* and then toward her first and second indifference curves and beyond as lesser options. Any point at the other side of point C is not even an option.

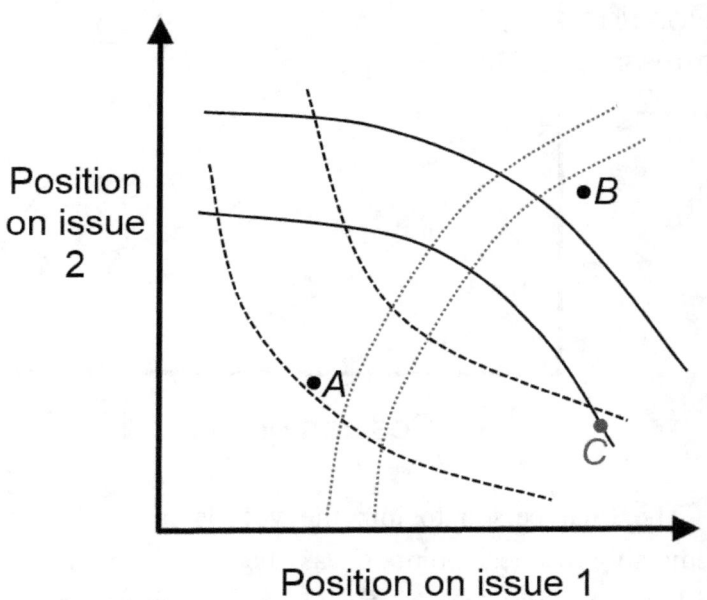

Position on issue 1

With the three voters all included in the above diagram and with their preferences visible some predictions can be made over how each would vote, along with the position on the two-dimensional issue that would win out here, if the three voters' ideal points A, B and C were the only ones considered.

Person A prefers point A to other alternatives, and point C is preferred to point B. This is clear as point C lies on person A's first solid indifference curve, while point B is beyond the second solid curve.

Person B would choose point B as his ideal, and point C would be chosen over point A. Although points C and A are both between person B's first and second dashed indifference curves, point C is closer to the first curve while point A is close to the less desired second dashed indifference curve in the distance.

Person C prefers point C as her ideal, and would choose point B over point A as the former is within her first dotted indifference curve, while the latter point A lies beyond her second indifference curve.

Putting all of this information together allows for voting preferences to be compared for each of the three voters, where the > sign represents a preference for one option over another:

$$\text{Voter A preferences: } A > C > B$$
$$\text{Voter B preferences: } B > C > A$$
$$\text{Voter C preferences: } C > B > A$$

With these voter preferences a prediction can be made as to which of the three proposals would win out. This prediction depends on these three points being the only proposals put forward in a two-dimensional policy issue vote, and there being equal numbers of each voter type (for example just one of each voter here). If all three positions were presented at the same time then the first round would see a draw, with one vote each for ideal points *A*, *B* and *C* from the corresponding voters. This is clear by simply looking down the first column of preferences:

A
B
C

Because of the tie in the first round of voting, the voters' second choice for the three positions would come into play. This is the second column in the set of voter preferences above. There are two votes for point *C*, one for *B* and none for *A*. Therefore point *C* would win the contest if the three options were presented to voters at the same time:

C
C
B

But perhaps the options are not presented to voters at the same time, and are instead presented in binary contest where the winning proposal in the first round advances to the next round. It could be choice *A* vs. *B* and the vote winner faces off against option *C*, or *A* vs. *C* with the winner facing choice *B*, or *B* vs. *C* and the more popular choice faces *A* in the final vote. The voters will always select the option that shares their name first, whether *A*, *B* or *C*, and so the only question is their preference between the other two options. Voters A and B both have choice *C* as a second preference, while voter C has position *B* as a second choice.

In a binary vote of positions *A* vs. *B* voter C has the deciding say and she would choose to see *B* win the contest. If it was choice *A* vs. *C* then voter B decides who wins, and he would select position *C*. And if it was a contest between options *B* vs. *C* then voter A decides who wins, and he would select *C* as the winner. The result is that the position *C* will win any binary vote it takes part in, irrespective of the order of the voting procedure.

A proposal that cannot lose against any alternative that it faces is called a Condorcet Winner. With *C* winning any theoretical binary election or referendum, as well as an election containing all three proposals at once, *C* is the Condorcet Winner here and it is the proposal that will always come to pass under majority voting.

When the only positions to vote for are ideal points then there may be a Condorcet Winner, as the ideal points turn a multidimensional issue into a single one (i.e. this is someone's single ideal, and it's been met), and preferences of course will be clear and single-peaked. This gives the characteristics required for the median position to never lose, and as the image containing all of the proposals shows, proposal C sits between the other two and looks like the median position.

7.2 Condorcet Voting Cycles

If all of the proposals put forward are ideal points then there may be a Condorcet Winner, but in reality not every voter will always be able to find their ideal point among the proposals put forward. Some voters may have preferences and positions on issues that are not met by the popular vote, and find only second best options on offer when it comes to the ballot box.

Imagine the election/referendum in the last section is modified, and the ideal point A of voter A is removed from the contest and replaced by a less preferred proposal not of his choosing, point a. This point lies to the north-east of the crossed out ideal point A in the following diagram.

The change to the proposals available to vote on has an effect on the preference ordering of the three voters. For voter A the new non-ideal point a simply replaces A as the preferred option, and while proposal C lies on his first indifference curve proposal a sits inside it. For voter C the new a proposal also takes the former position in preferences held by proposal A, and it's the new least favoured option. But for voter B the new proposal a is far more desirable than position A ever was, and as the diagram shows it replaces proposal C as his second best option, with a sitting inside his first indifference curve as position C lies outside it.

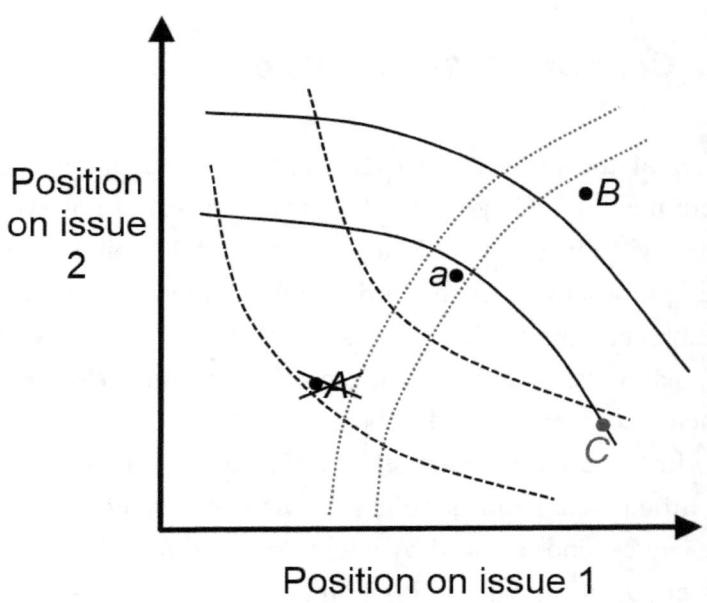

Position on issue 2

Position on issue 1

The new voter preferences are therefore as follows, with the > sign again representing a greater preference:

Voter A preferences: $a > C > B$
Voter B preferences: $B > a > C$
Voter C preferences: $C > B > a$

A quick glance at the columns of first, second and third preferences shows that there won't be a winning proposal here if all of the positions are up for vote together. The first column of preferences has one vote each for *a*, *B*, and *C*, and this pattern is also seen in the secondary preference second column, and the last choice third column. Because there is no winner when there is an open election/referendum with all three positions, the only solution is for there to be binary votes between the three proposals in turn.

If the first binary vote on proposals was a vs. B with the winner to face C, then voter C has the deciding vote in the first round and her preferences suggest she'd vote for B over a. That sees proposal B vs. C in the final round and the deciding vote is held by voter A, and his preferences suggest position C as the winner.

If the first binary vote was proposal B vs. C with the winner to face a, then voter A has the first deciding vote and would vote for C as just noted. That means C vs. a in the final round, and voter B decides it by selecting preferred choice a.

And finally if the first binary vote was between proposals a and C then voter B decides it by selecting a as just noted, and the final vote of positions a vs. B should see voter C decide it by going for proposal B as shown above.

There is no Condorcet Winner with the preferences held here and the inclusion of a non-ideal position. As the three examples show, the two proposals that fought it out in the first round never win out overall, and it's the proposal that only enters in the final round that is ultimately chosen. That means that it's the newest proposal seen by voters that will come to pass here, and this is likely to lead to voting cycles, or Condorcet Cycles as they're also known. Whenever a new proposal is introduced it will beat the old ones, whether its ideal points B or C, or non-ideal point a, and each time a, B, or C is put forward again it will win once more. This absence of a clear preferred position creates uncertainty in voting, and that's a situation that can be exploited.

7.3 Playing the System

The example in the last section with non-ideal positions in an election or referendum has some big implications for the voting system, and it suggests that the deciding factor will be the order with which the positions are presented to the voters. This allows policy makers to potentially manipulate the system if they had a preferred choice of their own, perhaps something that is a little better for those who are in power than other options, and they could simply offer this proposal to voters last in a round of voting to see it more likely to win public approval.

More worryingly for the integrity of the voting system, the inclusion of non-ideal points and the resulting voting cycles open up the possibility of tactical voting, where voters becomes disillusioned with the options offered by the system or see an opportunity to play it as best they can. Keeping the same three voters as used throughout this chapter, and including only B and C as ideal points along with non-ideal point a as in the last section, a two-stage binary vote can be tried again. But this time one of the voters is cynical, and they know that the proposal joining the fray in the last round will triumph, but their own preferred proposal will enter in the first round, on the request of policy makers.

The first round of the vote is position a vs. B, with the winner facing C. When voters weren't disillusioned voter A chose option a while voter B went for B, giving voter C the deciding vote which went for B. The B vs. C final round binary contest gave voter A the power to decide it, and his preference made C the winning proposal. But now that this choice has come around again voter B has a plan, and although he foresees his least favoured option of proposal C winning the overall vote he's going to try and change that, as he knows about the voting cycles and that none of the proposals are unbeatable.

Voter B can look ahead and see that in a final round between proposals a and C he would have the deciding say, as voters A and C would predictably divide their votes. He could then avoid his least preferred proposal C and instead push his second choice of a as overall winner. So the first round of position a vs. position B sees one person lie about their preference, and the dishonest voter B goes against his ideal point and instead votes for a to proceed. As voter A picks a too it gets two votes out of three and proceeds to the final round, where voter B sees it win out over proposal C just as he'd planned.

All it took for the voting system to break down was the inclusion of just one non-ideal point in place of an ideal one. That removed position C as the natural winner, led to the possibility of government manipulation as the vote winner would be the proposal presented last to voters, and saw a voter voting against his ideal position to manipulate the outcome of the contest. The uncertainty the non-ideal point creates in choice led to spiralling uncertainty throughout the entire voting system.

To maintain a working voting system in a democratic society the solution appears to be restricting voting to ideal points only. This may be difficult under direct democracy where citizens vote on very specific proposals with referendums, but it is possible in indirect democracy where representative candidates are selected with elections. The representative candidates would never implement anything other than their own ideal positions if elected, and the only question would be whether voters could find adequate representatives to choose from to follow their ideal points.

The search for representatives can explain the popularity of identity politics, where voters look for people from the same background who they assume have the same ideal points, only with more power to implement them. It also shows why it's often the more charismatic politicians and not those with well thought out policies that are chosen in elections. A good policy is still unlikely to be a truly ideal point, while a charismatic candidate could make everyone feel he's 'one of them' and with the same ideal points.

Bibliography

Abdellaoui, M., Bleichrodt, H. and L'Haridon, O. (2008) A Tractable Method to Measure Utility and Loss Aversion under Prospect Theory, *The Journal of Risk and Uncertainty*, 36, p.245-66.

Allais, M. (1953) Le Comportement de l'Homme Rationnel devant le Risque: Critique des Postulats et Axiomes de l'Ecole Americaine, *Econometrica*, 21, p.503-46.

Bernoulli, D. (1954) Exposition of a New Theory on the Measurement of Risk, *Econometrica*, 22, p.23-36.

Kahneman, D. and Tversky, A. (1979) Prospect theory: An Analysis of Decision under Risk, *Econometrica*, 47, p.263-92.

Lichtenstein, S. and Slovic, P. (1971) Reversals of Preferences between Bids and Choices in Gambling Decisions, *The Journal of Experimental Psychology*, 89, p.46-55.

Loomes, G., Starmer, C. and Sugden, R. (2003) Do Anomalies Disappear in Repeated Markets?, *The Economic Journal*, 113, 486, p.C153-66.

Machina, M. (1987) Choice under Uncertainty: Problems Solved and Unsolved, *The Journal of Economic Perspectives*, 1, p.121-54.

Schmidt, U. and Hey, J. (2004) Are Preference Reversals Errors? An Experimental Investigation, *The Journal of Risk and Uncertainty*, 29, p.207-18.

Schmidt, U., Starmer, C. and Sugden, R. (2008) Third-generation prospect theory, *The Journal of Risk and Uncertainty*, 36, p.203-23.

Simon, H. (1959) Theories of Decision-Making in Economics and Behavioural Science, *The American Economic Review*, 49, p.253-83.

Starmer, C. (2000) Developments in Non-Expected Utility Theory: The Hunt for a Descriptive Theory of Choice under Risk, *Journal of Economic Literature*, 38, p.332-82.

Sugden, R. (2003) Reference-dependent subjective expected utility, *The Journal of Economic Theory*, 111, p.172-91.

Tversky, A. and Kahneman, D. (1992) Advances in Prospect Theory: Cumulative Representation of Uncertainty, *The Journal of Risk and Uncertainty*, 5, p.297-323.

Von Neumann, J. and Morgenstern, O. (2007) *Theory of Games and Economic Behaviour*, Princeton: Princeton University Press.

www.ingramcontent.com/pod-product-compliance
Lightning Source LLC
Chambersburg PA
CBHW051732170526
45167CB00002B/904